I0616120

SURVIVING THE
AI SHiFT

*A Strategic Guide for Educators and
Staff in the Artificial Intelligence Age*

Ayo Jones, M. Ed.

Noodle Knowledge
Publishing

Published by
Noodle Knowledge Publishing

Surviving the AI SHiFT

A Strategic Guide for Educators and Staff in the Artificial Intelligence Age

By Ayo Jones, M.Ed.

Published by Noodle Knowledge Publishing
An imprint of Noodle Nook LLC
Copyright © 2025, Noodle Nook LLC. All Rights Reserved.

For permissions requests, inquiries, or further information, please contact:
ajones@noodlenook.net or hello@survivingtheaishift.com
(Or mail to: Noodle Knowledge Publishing, PO Box 112, Oregon, WI 53575)

General Disclaimer:
This book is intended for informational and educational purposes only. While every effort has been made to ensure the accuracy and effectiveness of the information presented, the author and publisher are not engaged in rendering legal, financial, or other professional services. This content is not a substitute for professional advice. If expert assistance is required, the services of a competent professional should be sought. The application of the principles and strategies outlined in this book will vary based on individual and organizational circumstances.

Printed in the United States of America.

ISBN: 979-8-218-71811-4

Dedication

To my family, for their unwavering support as I champion human ingenuity in the Intelligence Age. To my friends, for their believing and cheerleading. To my staff, for letting me stay in my zone of genius.

To every educator and staff member navigating this exhilarating, yet often challenging, AI-powered future. May this guide help you shift with confidence so you can continue to shape the incredible minds in your care.

Table of Contents

The Ground Beneath Us Is Moving

> *"The illiterate of the 21st century will not be those who cannot read and write, but those who cannot learn, unlearn, and relearn." - Alvin Toffler*

The sense that things are changing faster than ever before isn't just a feeling... It's reality. We are witnessing the dawn of what Eric Schmidt, former CEO of Google, describes as an AI revolution that is currently *underhyped*, not overhyped. This may be the revolution of all revolutions, requiring a profound shift in how we think, work, learn, and live, and demanding a new level of AI literacy from us all.

The ground beneath us is shifting faster than most of us are prepared for.

For decades, educators operated in a world of information. Our job was to curate information, deliver information, and assess information mastery. The **Information Age** is over. We've entered the **Intelligence Age,** where machines can generate, summarize, analyze, and even personalize information in seconds.

And that changes everything.

This shift is technical, but it is also existential. It challenges our understanding of knowledge, intelligence, and even what it means to be human. This rapid shift of AI capabilities is outpacing our collective ability to adapt, leaving many of us scrambling to catch up. Developing foundational AI literacy is not optional... It's the critical first step in keeping pace and regaining our footing.

This book serves as your strategic guide for navigating the most profound shift in the educational landscape we've ever seen.

Why This Book, and Why Now?

I wrote *Surviving the AI Shift* for every adult in our educational institutions - from the superintendent to the secretary, from the teacher to the counselor - with the intent of empowering you all to thrive in the AI Age. Here's the thing... We are all trying to survive this shift, but adults in education don't have the luxury of waiting to see where this goes or of making critical mistakes. The readiness of our institutions, the effectiveness of our staff, and the very success of our students are at stake. Our professional practices and ways of working are built for a past era, but we're still expected to navigate a future we can't fully see.

When I stand at the edge of this divide between the practices of the past and the technology of the future, I look down and see how wide this gap is. As a professional development strategist, tech integration expert, and someone who's spent over two decades in education, I can see how much our field has changed. Most of us have known our systems needed an overhaul for a while. In this moment, I can tell you one thing for certain: *Our old ways of thinking and interacting with technology will no longer work in this new Intelligence Age.* And the longer we wait to change, the more detrimental it will be for everyone. Remember those pre-flight safety briefings? When the plane hits turbulence and the cabin pressure drops, they tell you to put on your own oxygen mask before helping anyone else. Folks, it's time to mask up!

We have to rethink our professional practice in education. That means rethinking how we:

- Equip all adult staff to think strategically, question assumptions, and create innovative solutions in a world where AI is ubiquitous.
- Design professional development that leans into the process of adult learning and adaptation as much, if not more, than fixed products.
- Embrace AI tools as strategic enablers for professional practice and not just efficiency gains.

We all have to shift. And this book walks you through **The SHIFT System™**, a powerful methodology to recenter your professional practice and personal mindset around the skills that matter most in this new reality. Each element – **See, Humanize, Iterate, Frame, and Think** – is a strategy that you can incorporate into your daily workflows, professional development, or leadership initiatives, even if you have no tech or AI access!

To take your learning further, The SHIFT System™ AI Jumpstart offers a complete online program. It includes a hands-on Team & Leader's Playbook plus video module to get you started. You can begin shifting your institution into the AI Age today.

Learn more at **SurvivingtheAIShift.com/Jumpstart**

It's time to make the **SHIFT** into the Intelligence Age.

CHAPTER 1

Understanding
the Pace of Change

*"What technology rocked your
world when you were a kid?"*

The Technology That Shaped Us

When I was small, we didn't have remotes... *We were* the
remotes. My parents would call from the couch, and one of us
would run over to twist the knob, change the channel, and adjust
the antenna. Not that there were many options. A handful of
channels and a handful of hours was all network TV offered.
If you wanted cartoons, you had to get up early on Saturday
morning and park yourself in front of the TV before they were
gone until next week.

When I was a little older, something magical happened. We got a personal computer. It felt futuristic! This one machine in the house with floppy disks, pixelated graphics, and green text on a black screen felt like my own personal NASA command center. I remember sitting at that computer for hours, completely immersed in *Oregon Trail.* Somehow, I managed to avoid dysentery and make it out West, which felt like the greatest achievement of my life at the time. *How much better could technology possibly get?*

It turns out... a lot better.

And faster. Much, much faster.

Every Generation Has a Defining Technology

Every generation has a signature piece of technology that shifts how they communicate, gather information, and relate to the world. That one tool becomes a marker for their coming of age. And the rate at which that tool spreads is a signal of how fast the world is changing.

Silent Generation: The Age of the Radio
Born: 1928-1945. High School Graduation: 1946-1963.
Retire: 1993-2010.

The Silent Generation grew up during a time of global uncertainty with the Great Depression, World War II, and the years that followed. In this era of ration stamps and war bonds, the radio was a lifeline. It connected people to the outside world, offering news, presidential fireside chats, and entertainment in a time when printed media could not keep up with real-time updates.

For many families, it was a communal experience... one radio in the house, turned on during dinner or before bed.

The adoption of radio took nearly 20 to 25 years to become a household staple. That's because early radios were expensive, the infrastructure was still developing, and content was limited. But once it hit its stride, the radio became the center of the living room, and the first taste of mass media as we now know it.

Baby Boomers: The Television Generation
Born: 1946-1964. High School Graduation: 1964-1982.
Retire: 2011-2029.

Following the war, the Baby Boom era ushered in prosperity, suburbia, and a booming birth rate. This generation came of age alongside a new centerpiece in the American home: the television. While radios brought sound, TV brought sight... and with it, an entirely new way of experiencing the world. From civil rights marches to moon landings to Saturday morning cartoons, television made national events personal and visual.

Television took 15 to 20 years to reach broad adoption. Early sets were costly, and programming was limited to a few networks. But once color TV, sitcoms, and primetime news became common, families quickly reshaped their lives around screen time. The impact was profound. For the first time, generations were united by the same visual media moments at the same time.

Generation X: The Rise of the Personal Computer
Born: 1965-1980. *High School Graduation: 1983-1998.*
Retire: 2030-2045.

Generation X straddled the analog and digital divide. Born into a world of rotary phones and paper maps, they came of age as the personal computer entered homes and schools. While mainframe computers had been around for decades, the PC brought that power to individual users. For the first time, kids could word process, play games, or experiment with coding from a desk at home.

The PC was a game-changer for education. Word processing classes, computer literacy, and computer labs began appearing in schools. Tools like *Oregon Trail* and *Math Blaster* were early forms of edtech.

The personal computer took 10 to 20 years to achieve household penetration. Prices, operating systems, and hardware all evolved rapidly, but early adoption was still slow due to cost and technical barriers. Nonetheless, Gen X witnessed the beginning of a digital revolution that would redefine the workplace and the educational experience for adults and students alike.

Millennials: The Internet Revolution
Born: 1981-1996. *High School Graduation: 1999-2014.*
Retire: 2046-2061.

If the PC was powerful, the Internet was transformative. Millennials were the first generation to grow up with the ability to connect to the world from their homes. Dial-up connections may have been slow, but the access they granted to knowledge, communication, and entertainment was revolutionary. AOL chat rooms, Napster, early Google searches... These were formative experiences for a generation that began to see knowledge as borderless.

For education, this meant research moved beyond encyclopedias. Students could collaborate online, email teachers, and discover ideas beyond their textbooks. With the world's information so easily accessible, learning shifted from static to dynamic. For educators and administrators, it changed how they accessed professional resources, collaborated across schools, and engaged in new forms of professional development.

Internet adoption took less than 10 years to explode once broadband entered the scene. Unlike earlier technologies, the Internet became indispensable almost immediately. It laid the foundation for everything that came next, from social media to streaming to remote learning.

Generation Z: The Smartphone Generation
*Born: 1997-2012. **High School Graduation:** 2015-2030.*
Retire: 2062-2077.

Gen Z never knew a world without the Internet... and for many, they've never known one without smartphones. These devices collapsed the world into a pocket-sized screen. Text, video, GPS, music, camera, calculator; it was all right there, 24/7. Smartphones created an "always on" generation, deeply attuned to digital communication, visual media, and constant connectivity.

Then came the pandemic. School closures pushed millions of Gen Z students into remote learning almost overnight. Zoom replaced classrooms, and learning shifted to Chromebooks and video calls. For many, it was the first time education felt fully digital, and the cracks showed. While some students thrived with flexibility, others fell behind. The system wasn't ready. Devices and Wi-Fi weren't evenly distributed. Teachers lacked training. Parents weren't prepared to fill the gap. For instructional and support staff, the pandemic dramatically shifted our daily work, requiring rapid adoption of new digital tools for instruction, communication, and administration, often with insufficient training or support.

The impact on education was complex. Yes, mobile access to information and learning tools skyrocketed... but so did constant interruptions and digital fatigue. For Gen Z, the classroom could live in their hands, but so could every distraction imaginable. For adults in education, the digital shift meant new demands on our own literacy and adaptability.

Smartphones reached mass adoption in about eight years, making them one of the fastest spreading technologies in human history at that time. And the pandemic only deepened their importance, as both a tool and a crutch in the educational experience for everyone.

Generation Alpha: The Voice-Activated World
Born: 2013-2024. High School Graduation: 2031-2042. Retire: 2078-2089.

Gen Alpha was born into a world where yelling "Alexa" or "Hey Siri" gets a response. Smart devices are part of the furniture:

interconnected, voice-controlled, and intelligent. These children are growing up talking to machines that talk back. They swipe before they write and stream before they read.

Their early schooling years were shaped by an unprecedented disruption: the COVID-19 pandemic. Some started kindergarten on Zoom. Others entered first grade without ever having stepped foot in a classroom. The pandemic became a tech stress test... and it changed everything. While smart tools helped keep learning afloat, it became painfully clear that our education systems weren't designed for this kind of digital-first reality. Gaps in social development, inconsistent learning experiences, and increased screen dependence have become long-term issues. For adults in education, adapting to serve this generation meant a rapid overhaul of our own digital skills, workflows, and communication methods.

In classrooms, Gen Alpha is now being introduced to a world of adaptive learning platforms, AI-powered personalization, and hands-free tech. But their foundational experiences with education were shaped by uncertainty, and that will color their learning preferences and expectations for years to come. This profoundly and directly impacts how we all have to approach our professional development and daily work.

Adoption time for smart tech? About five years. That includes both device penetration and behavior normalization. Voice-first interaction is not futuristic for Gen Alpha... It's formative. And for many, the pandemic cemented that digital-first education is normal, even if imperfect.

Generation Beta (2025–2040): The AI Generation
Born: 2025-2039. **High School Graduation:** 2043-2057. **Retire:** 2090-2104.

This is where things get wild.

Generation Beta will grow up with AI and be profoundly shaped by it. From AI tutors to personal robotics to virtual reality adoption, this generation is entering the world as intelligent machines become integrated into nearly every facet of life. And unlike any generation before them, they won't remember a time before it.

Artificial intelligence didn't take 20 years to go mainstream. It didn't even take five.

AI tools like ChatGPT reached 100 million users in 3 to 5 months. That's not a curve… that's a cliff. The adoption rate has shattered every historical precedent, with reverberating implications for how we teach, learn, and live. For adults in education, this means an immediate and profound shift in our own professional practices, digital literacy, and adaptive capabilities. The pace of this change has been exponential. And Generation Beta is the first to be born into that velocity.

Generational Technology Adoptions
For the Silent Generation, it was the radio, adopted over the course of 20 to 25 years.

Baby Boomers were shaped by the rise of television, which took 15 to 20 years to reach widespread use.

Gen X came of age during the dawn of the personal computer, adopted over 10 to 20 years.

Millennials experienced the explosion of the Internet, which reached mass adoption in under 10 years.

Gen Z grew up with smartphones, ubiquitous in less than 8 years.

Gen Alpha lives in a world of voice-activated assistants and seamlessly connected devices, adopted in under 5 years.

Generation Beta, starting in 2025, won't ever know a life before AI, a technology that reached millions in under 5 months.

The adoption rate tells the story of this accelerating trend, where each generation's defining technology became part of everyday life in a shorter and shorter time span:

Generation	Born	HS Grad	Retirement	Defining Tech	Adoption Time
Silent	1928–1945	1946–1963	1993-2010	Radio	20–25 years
Baby Boomers	1946–1964	1964–1982	2011-2029	Television	15–20 years
Gen X	1965–1980	1983–1998	2030-2045	Personal Computer	10–20 years
Millennials	1981–1996	1999–2014	2046-2061	Internet	<10 years
Gen Z	1997–2012	2015–2030	2062-2077	Smartphone	~8 years
Gen Alpha	2013–2024	2031–2042	2078-2089	Voice Tech + Connectivity	~5 years
Gen Beta	2025–2039	2043–2057	2090-2104	Artificial Intelligence	3–5 months

For nearly a century, society had time to adjust. New tools emerged, but they took years, sometimes decades, to reach mass adoption. As adults in education, we had space to shift our approach, adapt our instruction, and figure out how to teach with new tools or work around them.

But AI didn't arrive slowly.
It appeared, matured, and spread almost overnight.
Tools like ChatGPT reached 100 million users in just two months.

That kind of mass adoption has never happened before.
Not with radio. Not with television. Not even with smartphones.

We've entered a moment of exponential acceleration.
The pace of change is no longer steady. It's vertical.

And that means professional development in education can't wait. We have to shift!

CHAPTER 2

What Are We Doing Here?

I was born, raised, and educated in the United States. For most of my adult life, that's where I worked and where I taught. But everything changed in 2020. Just as the world started to open back up after the pandemic, my family and I packed up and moved to Ghana, West Africa. Like so many others during that global shift, we chose something new.

Since then, my kids have been homeschooled using an online curriculum. It's been flexible, and it fits our lifestyle. But that doesn't mean it always makes sense.

In the past year or two, my teenage son, squarely in Gen Z, has said more than once that school feels like a waste of time. And honestly, I get it.

Most of what schools teach wasn't built for the world he's entering. The pace of technological change has outpaced the structure of traditional education. As an educational strategist, I know this has left many adults in education unprepared for the new demands of the Intelligence Age. As a parent, I've had to tell him the truth. A lot of this work may feel meaningless. We do it because we have to check certain boxes. But while we follow the system's requirements, I make sure he's also learning what matters most for his future, and what we, as adults, must master now.

He's studying prompt engineering. He's exploring how AI agents function and how they're changing work. He's reading about the ethics behind artificial intelligence and automation. Because that's where he's headed... That's the job market he'll graduate into and the reality every adult in our educational system is now navigating.

It's a strange thing to guide future generations through a landscape you know is being rewritten in real time, especially when you realize you need to adapt your own navigation skills first.

And it's not just my son's generation. It's *every* generation. We all need to shift our mindsets and practices.

The Sandwich Era

What does it mean to work in a field tasked with preparing students for the future when that future keeps evolving faster than ever before?

If we continue down the path where AI literacy means teachers use AI solely to create worksheets and essays faster, students use AI to answer or write those assignments, and then AI spits out grades and comments on them, *what is the point*? This cycle doesn't move our professional practice forward. It turns our work in education into a game of passing AI-generated content back and forth instead of building real skills.

We are living in a unique, often challenging, transformational period. On one side are the familiar, long-established practices and expectations of the Information Age that many of us rely upon. On the other, the new technologies and powerful tools of the Intelligence Age that change how we access information, solve problems, and communicate. This overlap creates a powerful pressure point in our schools, our departments, and in our classrooms. I call this the **Sandwich Era**: the intense squeeze we feel as our established ways of working are caught between these two rapidly diverging realities.

This sandwich creates a profound urgency in this moment. **Generation Beta is entering pre-kindergarten in just a few years**. These children have grown up with Alexa telling them bedtime stories, Siri playing their favorite songs, and videos that respond to their voices and touches. They won't need to learn how to use AI... it will already be second nature. Their expectations for personalization and responsiveness are already taking shape. Educational institutions must be ready to meet them where they are, and *you* must become AI literate to do that.

The reality is clear. The world we all face will be shaped by artificial intelligence, machine learning, robotics, and automation. Institutions will depend more on creativity and data fluency than on repetition and memorization.

You need skills to adapt, communicate clearly, think critically, and use new tools to create solutions. These tools will look different from what you use today, and the problems will be more complex.

The responsibility is yours. We must participate in professional learning that builds confidence in the tools you naturally use. You must intentionally prepare for a shift that is already underway.

Our institutions have faced change before, but never at this speed. There was time to adjust to radio. Decades for television. Time for the Internet and personal computers. That kind of gradual change is gone.

What lies ahead demands focus, strategy, and the willingness to evolve how you engage in professional practices and professional development.

What Technology Is Coming Next?

We're already seeing the early signals of what's coming... and it's moving fast. The next wave of technology is not futuristic. It's here, quietly entering beta, slowly becoming embedded in the systems around us. These innovations will absolutely shift how adults in education work, but, more importantly, they will redefine who gets to participate in shaping the future of our institutions.

Some of the most impactful tools coming into schools and districts in the next 5 years include:

1. Real-Time Language Translation and Transcription

AI-powered earbuds or wearables with embedded translation tools now allow educators, staff, and students to access conversations and collaborate in their home language instantly. This means:

- Teachers can communicate seamlessly with multilingual parents during conferences, fostering stronger home-school connections.
- Administrators can engage with all community members, ensuring access to vital information and services for all.
- Staff meetings and professional development sessions can become truly multilingual spaces where language is no longer a barrier to understanding or participation for any adult.

Instant translation supports easy access to information and builds cross-cultural collaboration across the entire school community. Built into wearable technology, like glasses, it delivers barrier-free access and use for everyone.

2. Multimodal AI that Goes Beyond Just Text

Next-generation AI tools can now process and generate text, images, video, audio, and code. Those same wearables have video capabilities that let AI "see" the world around you, including your lesson plans, student data, or facility diagrams. Processing these different forms of information is a shift allowing you to interact in the ways that work best for you.

- An educator who struggles with writing lengthy reports can speak their thoughts aloud and receive real-time support for a grant application or a curriculum overview.
- Staff can sketch new professional development concepts, record audio instructions for a new administrative process, or use dynamic visual models for complex data instead of relying on traditional written briefs.
- Complex district strategies can be explained through interactive visuals or narrated step-by-step demonstrations for all staff.

Ultimately, this blend of modalities serves to accelerate professional communication and personalize adult learning experiences for every educator and staff member.

3. Emotion-Aware Interfaces and Adaptive Learning Systems

Since these AI tools can see us and our environments, they can pick up on your non-verbal communication. By reading facial expressions, tone of voice, and engagement patterns, AI can adjust collaborative interfaces or provide tailored feedback automatically. These systems:

- Detect teacher burnout during high-stress periods, administrator fatigue in busy cycles, or staff confusion during complex professional development modules, and offer immediate, personalized support.
- Provide real-time prompts to re-focus attention during demanding tasks or suggest micro-breaks to prevent overwhelm for any adult learner.

- Offer personalized professional development pathways, alternative communication strategies, or even virtual coaching to re-engage adults in their learning.
- Help school leaders prioritize who needs support most in real time.

Rather than reacting to your needs, these tools help you respond as they arise and improve our overall well-being, professional effectiveness, and, more broadly, institutional outcomes.

4. Custom AI Agents as Study Partners

The idea that this technology is already here blows my mind! I used to wind up my cassette tapes with a pencil! Today, you can build your own AI tools to serve as professional research assistants, curriculum development partners, administrative efficiency agents, or personalized professional learning coaches. These personal AI agents:

- Support teachers with lesson planning, differentiating instruction, or organizing assessment data.
- Allow administrators to streamline scheduling, manage communication, or synthesize policy documents more effectively.
- Create consistent, on-demand support for any adult needing specific data, repeated practice on a new professional skill, or reminders for complex workflows.

Instead of using one-size-fits-all platforms, custom AI agents adapt to what you and your colleagues need.

5. Augmented Reality and Spatial Learning Tools

Augmented reality and spatial computing make abstract concepts tangible and collaborative. These tools already exist, but barriers to usage and adoption are quickly dropping. As that happens, more and more immersive professional environments and collaborative workspaces will be created, accessed using AR glasses and/or wearables. These tools:

- Bring 3D models of new school designs, complex budget visualizations, or detailed facility layouts directly into an administrator's workspace.
- Let staff teams "walk through" hypothetical emergency response scenarios, explore virtual professional development modules, or interact with student performance data visualizations as if they were physically present.
- Make hands-on training and remote expert guidance possible even for distributed district teams or in geographically disparate schools.

This technology brings depth, context, and innovation to complex educational challenges and professional development that might otherwise feel abstract.

Does this all sound like science fiction? It isn't. These tools are already here... and within five years, they'll be as normal as Google Docs and video calls.

The future of professional practice and professional development in education is personalized, multimodal, and powered by tools educators and staff don't have to type into... they can talk to, gesture toward, or simply collaborate alongside.

A Personal Note on Access and Opportunity

Living in a developing nation in Africa, I've seen firsthand just how wide the divide can be. Not just between countries, but within a single nation. A divide between urban and rural, connected and disconnected, resourced and left behind. Let's be clear:... This creates an opportunity divide, extending far beyond a simple digital one.

For years, access to quality education has depended on where you live, what language you speak, and whether your school has enough trained teachers or working devices. Too often, the answer to these questions determines whether a child gets a shot at a meaningful education or not.

But now, that's starting to change.

As AI tools become more affordable and more accessible, even in places where electricity can be inconsistent, we're seeing something remarkable: Access is starting to catch up. A child in a rural classroom with a single device and a strong idea can engage with the same level of content and support as a student in a high-tech lab. That matters.

When the barriers to learning begin to fall, the potential for innovation grows everywhere. And the next big idea, the one that changes everything, could come from someone who finally got access, not someone who always had it.

This is why what's coming next isn't just exciting. It's transformational.

Access for All

This new wave of technology may feel like one more thing added to an already full plate. But that only happens if we stay locked in the same instructional models we've always used. When we shift our perspective, we begin to see something different. Yes, these tools make teaching more efficient... But they also open doors that have long been closed.

For students with disabilities, language barriers, trauma histories, or limited access to resources, this is the most powerful moment of possibility we've seen in decades.

As a special educator and advocate, I see AI tools as assistive technology.

Now, on behalf of educators everywhere, I'm going to acknowledge the elephant in the room: When our students left for the spring break that never ended at the start of the pandemic in 2020, something happened. Things changed. The students all came back different... They came back special. And, honestly, so did we!

The pandemic shifted something in all of us. It left a lasting mark on how we live, teach, and learn. That shift has made one thing clear: Conversations about AI are truly conversations about access.

When thoughtfully implemented, AI becomes assistive technology for everyone. It can read aloud to students, summarize complex texts, scaffold assignments, organize thoughts, translate instructions, and adapt content in real time. It does what great teachers have always tried to do... it makes learning personal, responsive, and centered on the learner.

For years, we've talked about broad access as the goal. These tools make it possible. They level the playing field. Heck, they change the entire game!

To explore comprehensive AI literacy programs for your entire institution, including strategic planning and facilitator training, visit our website. Our programs help leadership teams prepare their workforce for the future of education.

Discover institutional solutions at:
SurvivingtheAIShift.com/Programs

Consider This...

These advances offer incredible promise, but they also demand a shift in how we think about education's role... not just in our classrooms, but in the broader world. To fully harness this potential, we have got to recognize the global landscape our students are entering and the disparities technology can help bridge.

While technology can feel like an added burden in well-resourced schools, in many parts of the world, it's the only path to high-quality learning. Where teachers are scarce or textbooks outdated, AI has the potential to deliver meaningful instruction, customized support, and real academic opportunity... even where little existed before.

Let's not forget: These tools aren't only showing up in schools, they're also defining the workforce. If our classrooms don't embrace this shift, we widen the gap between what students experience in school and what they'll encounter in life.

We have two choices:

- Adapt the system to reflect this new reality.
- Or continue training students for a world that no longer exists.

With both options, we have to consider that the workforce of tomorrow will truly be a global one, where competition for jobs is international!

As we stand on the brink of this new Intelligence Age, the choices before us have never been clearer. This new technology is not an add-on, a convenience, or optional. It is reshaping the very foundations of how we teach, learn, and connect. Embracing this change means more than adopting new tools; it requires a fundamental shift in mindset, practice, and purpose. The next chapter explores **The SHIFT System™** and how educators, systems, and communities must evolve to finally deliver our long-promised vision of accessible, high-quality education for all learners.

The future is not something that happens to us. It's not set in stone (or in this case, written in code). It's something we'll create together, starting now, by making the SHIFT.

CHAPTER 3

Making the SHIFT

> *"The time to adapt is now.*
> *We build the future of learning (and the future of*
> *mankind) with every choice we make."*

The world is transforming at a speed we have never witnessed. As we saw in Chapter 2, the pace of technological adoption has gone vertical, dropping from decades to mere months. We are living squarely in the **Sandwich Era**, caught between the familiar practices of the past and the powerful, rapidly evolving tools of the future. This is a fundamental change in how we work, lead, and guide learning, and what it means to be prepared.

For many educators, this reality feels big. The ground beneath us is shifting, and it can feel like we are about to be swept away. You can navigate this moment. This book shows you how to move beyond coping, leading you to succeed within this change.

The challenges before us are significant, but the path forward offers clear actions. You can learn to adapt to these changes, build new foundations, and reshape your professional practice in education for the Intelligence Age. This is where **The SHIFT System™** comes in. It is a practical, human-centered guide designed to help you recenter your professional development and daily work around the skills that truly matter in this new reality. Each element of **The SHIFT System™ – See, Humanize, Iterate, Frame, and Think –** offers clear strategies you can incorporate into your daily workflows, professional development initiatives, or leadership practice, even if you have no tech or AI access! It is time to make the SHIFT.

S: See (and Describe)

> *"The real voyage of discovery consists not in seeking new landscapes, but in having new eyes."*
> — *Marcel Proust*

When my son was three, he could spot an airplane in the sky before any of us adults even heard it. His eyes were no better than ours. His hearing was not sharper. His attention, however, was different. He was actively looking. He was seeing with intention.

We have moved away from this kind of active observation in the Information Age. We have become passive consumers of data. We scroll, we skim, we scan. We rarely stop to truly see.

In the Intelligence Age, observation is more than just a nice-to-have skill... It's survival.

Why Seeing Matters More Than Ever

Most people do not understand this about AI. It performs only as well as the instructions you give it. Those instructions depend entirely on how well you can observe and describe what you actually need.

Think about this. If I asked you to describe a sunset to someone who had never seen one, you would need to observe deeply. You would notice the way colors blend and shift. You would see how light reflects off clouds. You would observe the subtle changes in intensity and hue. The richer your observation, the more vivid your description.

AI operates in the same way. The depth of your observation directly impacts the quality of its output.

But here's what you, as an educator or staff member, needs to grasp. This is fundamentally about language. AI is built as a large language model. Nuances in words and syntax have a direct impact on outputs. Taking time to describe things with precision becomes an opportunity for enhanced professional communication and clarity in instructional or administrative context.

Consider the difference between these two prompts:

- "Create an image of a cat on a mat."
- "Create an image of an anthropomorphic orange tabby cat on a textured Arabian flying carpet."

The first prompt produces a generic result. The second, formed from careful observation and detailed description, creates something specific and compelling.

To make the input better, seeing and describing with depth of language is paramount. **SEE** develops the underlying vocabulary and communication skills to command AI rather than consume it.

The Lost Art of Deep Observation

We used to be better at this. Before smartphones, we noticed things. We watched people. We studied our environments. We paid attention to details because it was necessary.

Now, when you look up to people watch, you'll see everyone is in their own world, lost in an endless doom-scroll. Our attention is fractured. We have trained ourselves to multitask, to process information quickly, to move on to the next thing. But observation requires the opposite approach. It demands focus. It needs time. It benefits from stillness.

I see this in schools and departments all the time. I often watch staff members and teachers glance at a problem or new policy brief and immediately ask, "What am I supposed to do?" They don't take the time to see what is actually there. They're looking for the answer before they've even understood the question.

This is not their fault. We have trained ourselves, and our teams, to consume information rapidly rather than observe it deeply. In the Intelligence Age, that approach creates problems. Both human learning and AI collaboration require careful observation as their foundation.

What Deep Observation Looks Like

Real observation is active. It requires engaging with what you see, not just glancing at it. Here is how we can rebuild this skill:

The Observe-Reflect-Enquire Progression

This approach is not new, but it is more relevant now than ever. When you encounter anything (a new curriculum document, student data, a school policy, or an AI output) move through these three steps:

- **Observe:** What do you actually notice? Do not include what you think you are supposed to see. Do not include what you expect to see. What is literally there?
- **Reflect:** Based on what you have observed, what thoughts come to mind? What connections do you make? What patterns emerge?
- **Enquire:** What questions arise from your observations? What makes you curious? What would you like to know more about?

Unlike 'inquire,' which can suggest a search for a single, specific answer or a formal investigation, my use of 'Enquire' underscores the act of asking questions from genuine curiosity and a broader pursuit of knowledge.

Let me give you an example of this O-R-E progression. Last week, I asked a group of educators to observe an AI-generated image of a school common area during an event. Here is what happened:

- **Observe:** "There are 20 chairs arranged in rows." "Every person is looking forward." "The speaker is standing at a podium." "All the attendees have the same posture."
- **Reflect:** "This looks like a standard assembly setup." "Everyone seems very focused. It's a bit eerie." "The arrangement feels very formal and structured."
- **Enquire:** "Why are all the attendees sitting exactly the same way?" "Is this what AI thinks a typical school event

looks like?" "What images did AI train on to think this is 'normal' for an educational setting?" "What would happen if the seating was arranged differently?" "How would this environment feel to be in?" "How does this image's formality impact a sense of school community?"

That progression, from literal observation to thoughtful analysis to curious questioning, is the foundation of both critical thinking and effective AI collaboration.

Critical thinking has never been more important or harder to cultivate.

In an age where AI can generate convincing-sounding answers to any question, your ability to think critically about information, to evaluate, analyze, and question, is essential for navigating reality. A challenge every school leader knows: You cannot just tell your adult staff to "think critically." It is not a switch you flip. Critical thinking emerges from practiced habits of mind, and observation is where those habits begin. When we are intentional about building observation skills, we are laying the groundwork for the kind of thinking our educational institutions and you as an individual adult will need to succeed in an AI-saturated world.

N.O.T.I.C.E. and Note

Beyond the broader Observe-Reflect-Enquire progression, we can hone our observation skills even further by training ourselves to look for specific details. This approach, N.O.T.I.C.E. and Note, gives you a focused lens for sharpening your perception. It's about consciously scanning for certain types of information as you observe a situation or content.

Why does this matter for your team? In a world of overwhelming data and rapid-fire decisions, the ability to pinpoint critical details, spot anomalies, and recognize underlying patterns is a strategic advantage. It leads to clearer problem definition, more precise communication with AI, and ultimately, better educational outcomes.

As you observe, train yourself and your colleagues to notice these elements:

- **Nuances:** What subtle details, specific phrasing, or minor shifts stand out? This is about catching the fine print in a new district policy, the quiet feedback in a student survey, or the unspoken cues that might change the entire interpretation of a classroom dynamic.
- **Outliers:** What elements are surprising, unexpected, or contradict the prevailing narrative or data? Identifying outliers can reveal emerging student learning needs, overlooked professional development opportunities, or biases in your existing data sets. Don't dismiss them; investigate them.
- **Trends:** What patterns or repetitions do you identify across observations, data points, or project phases? Recognizing trends helps you predict future educational developments, identify best practices in instruction, or understand recurring challenges in staff workflows, whether in student behavior or administrative processes.
- **Inconsistencies:** Where do things not align, make sense, or seem to contradict each other? This critical eye helps you spot errors in reports, gaps in curriculum implementation, or misleading information, ensuring your team is working from a foundation of truth.

- **Critical Details:** What specific, significant elements appear important, even if small or initially overlooked? Often, the smallest piece of information... a keyword in a new state mandate, a specific data point in a vast student assessment report, a subtle change in a professional development module. These can be the most impactful.
- **Evolution:** How do things change or develop over time, if dynamic? Observing evolution helps you track professional growth, shifts in learning environments, or curriculum implementation journeys, ensuring your strategies remain agile and responsive to changing conditions.

Applying this process can change how you interact with information by guiding your initial perception towards critical elements that might otherwise go unseen. This deliberate practice sharpens your observational abilities and allows AI systems to be a more effective partner because you're more precise in your inputs and evaluations.

Using N.O.T.I.C.E. and Note with Your Team:

Don't dismiss this as theory. Use it as a practical team exercise. Integrate the N.O.T.I.C.E. checklist into your daily huddles, staff meetings, or professional learning communities. Here are a few ideas on how to integrate this for the biggest impact:

- **Shared Observation:** Display a piece of information relevant to your school or district team (e.g., a new student assessment report, parent feedback summary, an AI-generated curriculum overview, a visual representation of a school process flow).
- **Individual N.O.T.I.C.E.:** Give everyone 1-2 minutes of quiet time to apply the N.O.T.I.C.E. checklist to the information.

Encourage them to jot down their observations for each letter.

- **Collaborative Share-Out:** Go around the room (or virtual meeting) and ask each person to share one observation for one of the N.O.T.I.C.E. points. Facilitate a discussion around what surfaced. You'll be amazed at what different eyes perceive, and how quickly the collective understanding deepens.
- **Connect to Action:** Ask: "Based on what we've N.O.T.I.C.E.d, what questions do we have? What might be our next step? How does this change our understanding of the problem or opportunity in our school community?"

By making N.O.T.I.C.E. and Note a regular practice, you shift your team towards becoming more active, discerning observers, building a critical skill for navigating the vast and often ambiguous data landscape of the Intelligence Age in education.

Zoom In and Zoom Out

Effective observation also involves the ability to adjust your focus, much like a camera lens. This means purposefully toggling between the minute details and the broader context... zooming in and zooming out. Mastering this skill allows you to see both the individual components and how they fit into the larger picture.

This ability is especially crucial when working with AI. When you prompt an AI, the "lens" you provide (your understanding of the larger educational goal, the district's strategic plan, or the school's mission) directly impacts the relevance and quality of the smaller details it generates.

For instance, generating a curriculum overview without the district's overarching instructional goals in mind might lead to accurate content but irrelevant teaching insights. Similarly, asking for an image for a parent newsletter without providing clear school branding guidelines could result in a visually appealing but off-brand asset. The quality of your output isn't just about the specific request; it's about framing that request within the right context.

Practice toggling between detail and big picture:

- **Zoom In:** Focus on specific elements. What exact words are used? What precise colors appear? What specific behaviors do you notice?
- **Zoom Out:** Step back to see the whole. What is the overall message? What is the general pattern? How do all the pieces fit together? Consider the broader educational objective, the institutional culture, or the community context.

This ability to move fluidly between levels of detail is crucial for both understanding complex problems and communicating effectively with AI-powered tools. The consistent practice of zooming in and out develops a flexible mind. You, as an educator or staff member, gain a comprehensive understanding of any subject, moving beyond isolated facts to grasp overarching patterns and relationships. This skill is vital for making sense of complex information and for clearly communicating your needs and analyses to both human collaborators and AI systems.

The Dual Lens Exercise for Your Team:

This simple exercise helps prime your team to toggle between detail and the bigger picture, ensuring they approach AI prompts and educational challenges with the right context.

- **Choose a Visual:** Select a relevant visual from your educational setting (e.g., a student performance dashboard, a single slide from a new curriculum presentation, a piece of parent communication, or a diagram of a complex school process).
- **Zoom In First:** Ask your team to spend 30 seconds silently observing only the most minute details they can find. What specific numbers, phrases, colors, or icons jump out? What's happening in one small corner?
- **Zoom Out Next:** Then, ask them to spend another 30 seconds silently considering the overall purpose, context, or strategic goal of this visual. How does it fit into the broader school or district strategy? What's the main message it's trying to convey? Who is the intended audience and why?
- **Discuss the Connection:** Open a discussion: "What did you notice when you focused only on the details versus the big picture? How does understanding the small details impact your perception of the big picture, and vice versa? What new insights emerged when you held both lenses?"

By making this a regular practice, you help your team to **SEE** without getting lost in the weeds or floating too high in the clouds. Constantly connecting the two is a crucial skill for effective prompting and strategic decision-making in education.

Language Shapes Seeing

In this new age of intelligent tools, seeing clearly is only part of the work. The other part is being able to describe what you see (and what you want) with accuracy and style. That means words matter more than ever.

AI systems were built as large language models (LLMs), learning based on consuming vast amounts of text-based data. That means they were built to respond to language. The more precise your description, the more intentional the output.

If you ask for an image of "a girl walking in the rain," you'll get one thing. But if you ask for "a girl rendered in mosaic style, walking through stained glass puddles under a dark gothic sky," you'll get something else entirely. The difference wasn't in the idea. There was a clear inspiration in mind. The difference is in the language used to describe it.

This is where vocabulary becomes power. Beyond just naming objects, you and your colleagues need to develop a vocabulary that names style, tone, and feeling. Being able to say "mosaic" instead of "art," or "gritty realism" instead of "dark," gives your teams the ability to turn vision into results.

Style Is Part of Seeing

As we make this fundamental shift and start using AI more in our professional practice, you will start creating with AI tools. It's not enough for you to know what you want to create. You need the words to tell AI what that looks like.

What happens when you ask for parent communication copy written in the style of Shakespeare versus Aldous Huxley? One leans into poetic rhythm, metaphor, and Elizabethan drama. The other brings dystopia, irony, and philosophical undertones. In other words, the outputs are vastly different.

In video generation, asking for a dolly pan creates a moving side view. Asking for a cinematic zoom changes the emotional feel by pulling closer or further from the subject.

It may seem like these are all a bunch of technical terms, and they are. But this depth of vocabulary and contextual understanding is also the essential tool for innovation and creation. Words like these help shape how ideas take form. Working with AI requires precise language, and precision comes from practice. That includes being able to compare genres, play with artistic styles, describe tone, voice, or perspective, and name emotional feel, aesthetic mood, or cultural reference.

Descriptive Language Builds Creative Power

When you grow your vocabulary, you expand your ability to create. You learn to move from "make it cool" to "make it bold, high-contrast, and minimalist" or "make it inspired by brutalist architecture and scored like a Hans Zimmer trailer." That shift turns a vague idea into a clear directive.

This is also where we see real professional insight and strategic value take shape. An educator or staff member who can describe a concept clearly (visually, emotionally, or stylistically) is practicing high-level thinking. They are choosing words with care. They are crafting meaning on purpose.

Observation and language go hand in hand. The more you learn to see, the more you learn to describe. And the better you describe, the more precise and powerful your use of AI becomes. In the Intelligence Age, words are design tools. They are the building blocks of collaboration between humans and machines.

When you develop the ability to see and use descriptive language, you don't just get better AI outputs. You become better thinkers, better communicators, and better creators.

That's what the **"See"** part of **The SHIFT System™** is all about.

H: Humanize

"I'm sorry, Dave. I'm afraid I can't do that."
— HAL 9000, 2001: A Space Odyssey

When I was seven, I painted a portrait in art class. It wasn't amazing, but it was solid for a second grader. I picked great colors. I added texture. I even tried to give it dimension. But at the top of the page, in giant letters, I spelled my own last name wrong.

And not just a little wrong. I added three letters. It was like I got distracted midway and started over. In my defense, my last name was twelve letters long, but still.

When the teacher hung it up in the hallway with all the other portraits, I felt like it was a spotlight on my mistake. Every time I walked past it, I didn't see the painting. I saw the error. I saw what was wrong.

When it was finally returned to me, I didn't keep it. I didn't revise it. I ripped it to shreds.

I destroyed something I had made because I thought it was wrong.

Why Humanize?

Why tell this story? Because it's what happens when we disconnect people from creativity.

When we only focus on the flaws...
When we treat creation like a performance, not a process...
When we forget that the work we make is part of who we are...

We lose the very thing that makes professional practice impactful.

And that's why this part of **The SHIFT System™** matters so much.

Humanize is about protecting the part of the work that is still deeply personal. It's about reminding you (and our students) that we are not machines. And meaningful professional practice was never meant to be mechanical.

It was always meant to be human.

The truth is, you may not know how to fully tap into your creativity anymore. We've spent years standardizing our processes, trimming down our professional choices, and giving ourselves step-by-step instructions for everything, inadvertently stifling the very creativity we now desperately need. Now we're asking ourselves to be innovative... and lots of us don't know where to start.

That's not our fault... We've inadvertently trained creativity out of ourselves. But here's the good news: It's still there. It just needs space, support, and a shift in mindset.

That's where AI tools come in. Used intentionally, they can open doors instead of closing them. They can help us get unstuck, think bigger, and test new ideas without the fear of getting it "wrong." If we don't shift our mindset, we may start outsourcing our strategic thinking altogether. That's not what we want. Our goal is to amplify ideas, not automate them. That's why we have to humanize the way we use technology and equip ourselves in the educational workforce to work creatively, critically, and consciously with AI tools.

From Human Theory to Human Practice

If creativity is going to make a comeback in educational practices, it can't be theoretical. We have to create opportunities for you and your colleagues to practice creative thinking, experiment with ideas, and evaluate how they use AI in the process.

That means giving yourself chances to play... with language, with ideas, with outcomes. Here are three practical ways to do just that.

Tweak My Style

Sometimes, one word changes everything! We already covered how powerful language and vocabulary are when commanding AI tools. As you tap into our creativity, we have to play with how *style* and *word choice* shape output and meaning in generative tools. With this activity, let's explore style and creative variation by adjusting just one element of a prompt.

Instructions:

1. Start with a simple prompt (e.g., "Generate a lesson plan concept").
2. Choose a style or tone (e.g., "as a concise parent email," "in the voice of a seasoned principal," "like a formal district policy document").
3. Generate the AI response.
4. Swap out the style words and regenerate the prompt (e.g., "as a playful classroom announcement" or "like an urgent memo to the school board").
5. Compare the results. Reflect:
   - What changed in the tone, pacing, or details?
   - How did one word or phrase shift the creative direction?
   - Which version communicated your idea best and why?

This activity helps you see how AI interprets language, encourages creative exploration, and reinforces the power of intentional word choice. It shifts the focus from "what's the right prompt?" to "how can I shape the result through strategic style?" It also goes the extra step of showing how important humans are in the creative process. When we command with intention, we create without limits.

Don't have AI access in your school or department? That doesn't mean this activity isn't helpful. Draft a prompt and ask colleagues to change one word to alter the style. Get colleagues to pair up and explain why they made their target change and what impact they think it will have on the output. Again, the process and humanization is the goal.

The WonDAR Framework:
Wonder, Direct, Adapt, Review.

This simple framework guides you to remain an active thinker and creator in your interaction with AI.

- **Wonder:** Start with genuine curiosity.
  This is the human spark that begins any creative process. Encourage yourself to brainstorm broadly, explore different concepts, and ask open-ended questions before even thinking about AI. What are you truly curious about? What educational challenges do you want to solve? What new opportunities can you envision? This initial, unconstrained thinking is vital for shaping your inquiry.

- **Direct:** Guide AI's output with precision and intention.
  This means being deliberate about what you are asking AI to do and how you shape the prompt. Consider the specific educational context, the desired instructional or administrative deliverable format, and the tone you want to achieve. Directing AI effectively requires strong observation skills and careful language use, ensuring the tool works for your specific purpose.

- **Adapt:** Personalize what AI provides.
  This is where you inject your unique voice, pedagogical insight, or administrative expertise. You should change the voice, reformat the structure, or add your own examples and insights. Adapting means shaping AI's output into something that genuinely reflects your own thinking and meets your specific creative or strategic vision, making the work truly your own.

- **Review:** Critically evaluate AI's output.
  You should ask yourself if the content is accurate, relevant to your educational goals, and aligned with your initial intent. This step includes checking for biases, identifying missing information, and considering how AI's response might be misinterpreted. A robust review ensures the final product is reliable and meaningful.

This four-step process does two things. First, it shifts the relationship with AI from a mere content generator into a thought partner and collaborator. Secondly, as you start to internalize this process, it allows you to see your role in shaping AI's output and how AI tools reflect your intent and unique vision. This active involvement keeps the work human-driven.

AI-o-Meter

As we shift further into the Intelligence Age, distinguishing authentic human work from AI-generated content becomes increasingly difficult. This means you must develop the practice of disclosing your AI reliance and evaluating your level of tool dependence.

This practice serves a purpose beyond just following rules. It helps you understand your own professional contribution process, fosters transparency, and builds professional accountability. After using AI to complete a task or project, reflect on how much AI contributed compared to your final product.

3 Options for Self-Reporting:
Contribution Scale

Use a simple scale (for example, 0-5) to indicate AI's contribution. You select the option that best describes AI's role in your work:

- **0 = No AI used.** The work is entirely my own creation.
- **1 = AI for brainstorming initial ideas.** I used AI to generate concepts, but all content creation and structuring was human.
- **2 = AI for initial draft/support.** AI generated some content or helped with basic formatting, but I significantly edited, shaped, and expanded it.
  (e.g., AI drafted an initial parent communication, I added the school's specific tone and necessary disclaimers.)

- **3 = AI for substantial content.** AI generated a large portion of the content, but I provided specific guidance, refined the output, and added my own unique insights.
  (e.g., AI generated a lesson plan outline based on standards, I infused my pedagogical expertise and differentiated for student needs.)

- **4 = AI for primary content.** AI generated most of the content, with my role focusing primarily on minor edits, checks, and formatting.
  (e.g., AI created a draft for a new school policy, I reviewed for alignment with district values and legal compliance.)

- **5 = AI for nearly all content.** AI generated almost all the content, with minimal human intervention or unique contribution.

Circle the number that best describes AI's contribution to your chosen task:

0　　　1　　　2　　　3　　　4　　　5

Detailed Reflection Prompt

Provide a short reflection prompt to differentiate the human work from AI's. Use specific examples. For example: "What specific parts did AI help you do, and what did you do entirely yourself? How did your own thinking or creativity build upon, change, or direct AI's output?"

Process Breakdown

Start with a simple breakdown of the project into key stages (for example, research, outlining, drafting, revising, editing, formatting). For each stage, briefly describe AI's role and your own specific contributions. You can also couple this with the contribution scale to see how much content was AI-generated for each step in the project.

Self-reflection builds metacognition around AI's role in the professional creation process. It cultivates awareness of how and why you use the tool and helps you to see how much control you are relinquishing as part of your creation process.

Your Free AI-o-Meter Tool

To apply the AI-o-Meter immediately in your professional practice, download your free printable guide. This tool helps you consistently evaluate AI's contribution and foster transparency in your work.

Get your free AI-o-Meter at:
SurvivingtheAIShift.com/AIOMeter

A Human-Centered Shift

As we shift into this new era of professional practice in education, we have to keep the human element front and center. That means equipping ourselves in the educational workforce on how to stay engaged in the creative process, be active with AI tools, and not give away our strategic thinking for the sake of automation.

And, to be clear, creativity is a critical skill! It will define success in the modern workplace, right along with communication and strategic sequencing skills. As automation grows, uniquely human capacities, like creative thinking, divergent problem-solving, and meaningful storytelling, become even more essential.

In this Sandwich Era moment, we are all caught between the systems we've known and the tools that are emerging. Most of us will have to go back and re-ignite human creativity within ourselves and our staff. That means creating space for purposeful play, experimentation, and exploration with AI tools. When we do that, we cultivate a mindset of seeing AI as a tool, but humans as the ones who think, feel, imagine, and lead.

Years after I tore up that painting as a seven-year-old, convinced it was fundamentally flawed, I came to realize a profound truth: Our human creations, even with their imperfections, carry a unique value that AI cannot replicate. That moment reminds me that our unique perspectives, our willingness to try, and our imperfect human touch are what truly matter. Yes, AI can replicate a lot of what we create. It can even simulate connection. What it can't do on its own is prioritize humanity, the genuine spark, and the messy process. That part is still on us. We have to be mindful about keeping people at the center, making space for genuine interaction, and designing with care so that the tech serves the humans... not the other way around.

That's what the **Humanize** piece of **The SHIFT System™** is all about.

I: Iterate

> *"In the age of AI, the first answer is never the final answer."*

When AI image generation first hit the scene, I was hooked. I had seen what people were making with tools like Midjourney, and I was amazed. The images were vivid, complex, and creative. Some looked like professional concept art straight out of a movie pitch.

So, I sat down to make my masterpiece.

And when I tell you it was worse than that painting I tore up at age seven... it was.

It was generic. Flat. Basic. The proportions were off, the colors were weird, and it just didn't *feel* right. I chalked it up to hype. "I guess these tools aren't as good as people say," I thought. And I moved on. I didn't try again.

What I didn't realize at the time was that **iteration is part of the creation process.**

I was still thinking like it was the Information Age. Ask a question, get an answer. But these new tools weren't search engines. They were *collaborators.* The kind of collaborator that needs feedback, refining, rewording. They work best through *interaction.*

And I hadn't shifted to that mindset yet.

The Power of Iteration

In the Intelligence Age, iteration (*the process of refining, repeating, and improving an idea or action based on feedback*) is a core competency.

AI tools rarely give you a perfect result on the first try. They give you a starting point. That means shifting our approach from AI consumers to AI collaborators. This shift means iteration is more than just part of the process, it's the heart of it.

Moving past the initial output represents true thinking. That's what makes iteration such a critical skill. It's not about getting it right the first time. It's about knowing how to improve it over time.

Learning how to refine, rethink, and reframe in response to AI output is one of the most valuable skills you can cultivate in this shift. Why? Because:

- It produces stronger results than first attempts.
- It builds resilience and reduces fear of failure.
- It mirrors how innovation works in the real world.
- It aligns with how generative tools function... through loops, not one-offs.

For years, many of us operated with the goal of "getting it right." But in this new era, *getting it better* is the new benchmark. That shift requires iteration.

The BID & ReBID Cycle

Effective iteration with AI works best when you have a structure to guide your thinking. The process of refining and reworking is a major shift that needs a framework.

That's where the BID & ReBID Cycle comes in.

Adults in education need to move beyond single-shot prompting. You need to learn how to build better questions, evaluate what AI gives back, and try again with intention. These questioning and revision skills are the new critical educational asset.

Working with AI is a lot like placing a bid at an auction. You start with your best offer... your clearest question or command. But if the return isn't quite right, you don't walk away. You rebid. You improve your offer to get a better result.

This is how the BID & ReBID Cycle cultivates the skill of thinking iteratively and communicating more clearly.

The Initial BID Process

This is the first step of the process. It helps you shape your curiosity into strong starting points.

B – Brainstorm

Begin by generating a wide range of questions related to a topic. There are no wrong ideas here. The goal is open thinking and curiosity. Let yourself wonder and enquire.

I – Identify
Next, sort through the brainstormed ideas, individually or as a team. Choose questions with the most potential, looking for ones that are open-ended, interesting, or connected to a purpose. Peer conversations often help clarify which ones are worth exploring.

D – Draft
Finally, refine your selected question to make it clear and specific. This is where skills from the **SEE** phase come into play. Adjust wording, add needed detail, and polish the question into a strong initial prompt.

This **BID** process is especially powerful when done collaboratively. It encourages conversation, careful thinking, and intentional question design. Plus, it builds communication and negotiation skills in an authentic way, pushing human connection. And once you send that first prompt to AI, the response becomes the starting point for the next round of thinking.

The ReBID Process
The ReBID process is where iteration becomes visible. We respond to what AI has generated by evaluating the results and deciding how to improve them.

RE – Review & Evaluate
Read AI output carefully (review). A specific tool to evaluate that output (the CLEAR Check) is introduced after the example.

B – Brainstorm
Generate new ways to approach the question based on what was missing or unclear in the output. You don't just start over, you build from what you've got.

I – Identify
Select the most promising revision idea. This might mean changing the tone, narrowing the focus, or adding context to guide AI better.

D – Draft
Finally, write a revised question or prompt, shaped by everything you've learned from the first round.

An Example of BID and ReBID in Action
Let's look at what this process might sound like in a real educational scenario, using a topic that often impacts school staff, students, and administration.

School/District Topic: Evaluating the potential impact of a new bell schedule on staff collaboration and student well-being.

Initial Prompt (BID):
"What are the pros and cons of implementing a new bell schedule for our middle school?"

This prompt was created after brainstorming several questions about improving school operations. The team BID by brainstorming lots of ideas and questions, identifying the most relevant one for their school community, and drafting it with a focus on clarity and purpose.

AI Output:
The response lists several generic points:

- Pro: More sleep for students
- Pro: Longer class periods
- Con: Challenges for after-school activities
- Con: Difficulty for transportation

It's accurate, but surface-level. It lacks specifics about middle school staff's unique challenges and doesn't include recent educational research or nuanced perspectives on different learning models. It also reads like a basic list, not a strategic analysis for an educational institution.

RE – Review & Evaluate:
The team reflects:

- "The answer didn't go deep enough for our institutional needs."
- "It repeated points we already knew."
- "It didn't feel focused enough on staff implications and didn't help us understand the real impact on schools like ours."
  - Applying N.O.T.I.C.E.: "I noticed the 'pros' were all about student benefit, but didn't address staff workload or professional development time directly. The 'cons' were too generic, not specific to our school's operational workflows."

ReBID Process:
- **B – Brainstorm:**
  The team explores new ways to frame the prompt:
  - Could we narrow it to focus on staff collaboration opportunities?
  - Should we ask about schools that have already implemented this, and what were their professional development implications?
  - What about specific metrics like teacher planning time or student engagement during transitions?

- **I – Identify:**
  They decide to focus on both staff well-being and collaboration, specifically for middle school faculty.

- **D – Draft (New Prompt):**
  "What does current educational research and case studies say about the impact of a new bell schedule on middle school staff collaboration and well-being for U.S.-based schools, including specific metrics and reported outcomes from institutions that have already adopted this model?"

New AI Output:
This time, the response includes:

- Specific studies from schools that piloted new bell schedules
- Quotes from school principals and union representatives
- A breakdown of teacher planning time and cross-departmental meeting opportunities
- Notes about impacts on staff morale and efficacy within professional learning communities

The team reads the new output and feels like they've moved beyond generalizations and into useful, real-world educational insights.

What Changed?

This is the power of the **ReBID** step. Everyone on the team had a clear purpose. When AI's response didn't meet that purpose, they didn't just change the wording... They revisited their thinking. They looked for what was missing, adjusted their prompt, and tried again with intention.

This process doesn't always happen in one pass. Sometimes it takes two rounds. Sometimes it takes twenty. But each cycle sharpens both the inquiry and the response, moving closer to strategic clarity for our educational goals. And through that loop, they practiced essential skills: analysis, synthesis, evaluation, and communication.

Even without access to AI tools, this structure can still shape powerful staff collaboration habits. You can practice writing better questions, evaluating colleague responses, and improving your ideas through discussion. That kind of critical thinking builds resilience... and a bit of grit.

But how do we know when to ReBID, or what exactly to change? How do we move from "this isn't quite right" to specific, actionable improvements? That's where a simple, yet powerful, evaluation tool comes in: the **CLEAR Check**.

The CLEAR Check to Evaluate AI Output

One of the main ways you humanize what AI is generating is by analyzing and evaluating it. Not every AI output is good. Some are downright awful or flat-out lies. You need a clear way to figure out when something is off or incomplete. That's where the CLEAR Check comes in.

This was taken directly from my book *The PEACE Framework*, which provides a step-by-step framework to AI-powered inquiry-based learning. This translates directly to strategic problem-solving, curriculum analysis, and continuous operational improvement. As you review AI responses, ask yourself if it is **CLEAR:**

- **C – Complete:** Does this fully answer the question, or provide the comprehensive information needed for our educational objective?
- **L – Logical:** Does it make sense, align with known facts, and match what I was asking?
- **E – Explain:** Can the AI explain how it got this answer, or cite its sources if applicable? (This is key for trustworthiness.)
- **A – Adjust:** Could there be in a better format for our needs? A table? A list? A professional development brief? A curriculum comparison?
- **R – Reword:** Could I rephrase my question to get something better, or refine the prompt for a more targeted output?

This is a simple way for you to evaluate AI, instead of just accepting whatever is generated. It helps you reinforce the idea that using AI tools is different than using a vending machine. Our human and AI collaboration demands a back-and-forth process to get the best results.

Teaching and Practicing Iteration

Building the habit of iteration means designing professional development and institutional improvement cycles that center on progress and process, not perfection. Refinement should be expected, even celebrated, and never reserved only for those who "didn't get it right" the first time. It's about empowering everyone to get it better.

Here's how you develop this essential skill, embedding iteration into the very fabric of how you work:

Design Professional Development for Continuous Evolution.
Forget the "one-and-done" mentality for professional learning outcomes. True iterative thinking means building revision directly into the PD cycle. Leaders must structure professional development modules or school improvement initiatives so that multiple versions are explicitly required and encouraged. This creates a natural expectation that every submission is a stepping stone, not a final destination, providing built-in space to reflect and revise along the way. Your goal is to cultivate a professional learning environment where iteration is the default.

Promote Reflection as a Daily Habit.
After you interact with AI, or complete any task where iteration is key, build in dedicated moments to pause and think about the process. Use reflection to connect what you put in, what you got out, and what you learned strategically. This doesn't have to be a big formal review; it can happen through quick check-ins, brief digital journal entries, or simple exit questions after a task. Ask questions like:

- What would you change in your prompt to get better results next time?
- What surprised you about AI's response, or the outcome of your first attempt?
- What specific improvements did you see between your first and final version?

These small, consistent reflections help you truly see how your choices shaped the outcome, and more importantly, how you can continuously improve.

Give Feedback that Inspires the Next Step.
This is where your leadership really shines. Shift feedback away from simply pointing out corrections and instead, focus on open-ended prompts that encourage others to rethink and re-strategize your choices. This kind of guiding feedback empowers agency and ownership. Instead of saying "This lesson plan is missing X," try questions like:

- "What if we explored this from another pedagogical angle to deepen its impact?"
- "Is there a different word or tone we could try for our parent community that might resonate more?"
- "How could we revise this to better align with our school's mission?"

This is a hard but vital shift for most school leaders accustomed to fixed metrics or rigid curriculum implementation plans. In this new era, it's not just about the final product; it's profoundly about the process. Instead of telling adult staff what's wrong, invite them to explore what's possible.

Build Strategic Checkpoints into Complex Initiatives.
For longer or multi-stage projects, establish specific points to pause, reflect, and adjust your course. These planning meetings or professional learning communities are structured opportunities to apply the **BID & ReBID Cycle** on a larger scale. Teams can re-evaluate their overall approach, refine their core inquiries, and strategically redirect their efforts based on initial outcomes or changing student needs. Small course corrections early on help prevent teams from going too far down an unproductive path, saving significant time and resources in the long run.

Cultivate a Visible Culture of Revision.
Make iteration visible and valuable by sharing and celebrating examples of staff members who significantly improved their work through multiple versions. Highlight their journey and the process, not just the polished final product. Celebrate "draft number four" as a success story of perseverance and strategic thinking, not a sign of initial failure. The goal is to normalize trying again, refining, and making it clear that continuous revision is where real growth, innovation, and institutional improvement happen.

The "Worst-to-Best" Challenge.
This is a fun and highly effective exercise. Give your team a weak AI-generated output. You might intentionally start with a vague or poorly written prompt, or feed a simple command into an AI tool to produce a generic, uninspired result. The task for your team is to take that initial output and, through iterative prompting and human insight, transform it into something strong and strategically useful. They can use the ReBID

process to revise the original prompt, add critical educational context, refine the tone, restructure the inquiry, or guide the AI to respond in a far more meaningful way for a specific educational objective.

This challenge powerfully reinforces the core idea of iteration and dramatically shows how much stronger the results become when humans are actively and intelligently involved. It also highlights another key truth: There isn't just one "right" answer. There are many ways to make something better. It's very enlightening to see how different educators and staff approach improving the same piece, and it reminds everyone that the goal isn't uniformity... it's growth through process.

Iteration Speed Runs.
For a quick and dynamic practice, set a short time limit (e.g., 5-10 minutes) to revise or rework a prompt or an initial AI output multiple times. This rapid-fire practice helps you get comfortable making quick decisions, trying new angles, and refining under pressure. It mirrors real-world conditions where time is often limited, but strategic thinking and adaptability still matter.

The "Rapid Refine" Icebreaker for Your Team:
This is a fantastic way to incorporate iteration into a staff meeting or quick professional learning community check-in. It reinforces the idea that the first version is rarely the best, and rapid refinement leads to stronger results.

- **The Initial Idea (1-2 minutes):** Your facilitator will present a simple, common educational scenario. For example, "Draft a concise update on our Q3 student assessment data for a busy principal." Or "Summarize the key takeaway from yesterday's professional development session."

- **First Pass (1 minute):** Quickly jot down your initial draft of this update or summary. Don't overthink it, just get it down.
- **Refine for Clarity (30 seconds):** Now, quickly revise your draft, focusing only on making it clearer. What can be simplified? What jargon can be removed?
- **Refine for Impact (30 seconds):** Next, another rapid revision, this time focusing only on making it more impactful. What's the single most important message? How can it resonate more?
- **Share the Evolution:** Your facilitator will invite a few staff members to share their initial draft, then their "clarity" revision, and finally their "impact" revision.

You'll be amazed at how much a piece of communication can improve in just a few minutes of focused, iterative effort. This quick drill shows that even small, intentional adjustments can dramatically elevate output, reinforcing the power of iterative thinking in real time.

An Iterative Shift

Iteration is more than a revision strategy. It's a mindset. When you learn to reflect, refine, and rework, you shift from passive receivers to active creators. You begin to understand that growth doesn't happen in a single moment - it happens through a series of intentional steps. This shift builds confidence, deepens professional thinking, and prepares you for a world where the first idea is only the beginning.

That's what the **Iterate** piece of **The SHIFT System™** is all about.

F: Frame

> *"The quality of an AI's answer begins with the clarity of your question.*
> *Frame it well, and the possibilities are endless."*

"I dunno."

"Wait... what don't you get?"

"Huh?"

"Do you not understand the question? The directions? The process?"

"Um. Yes?"

Sigh.

This is an actual conversation I've had with more than one teenager... a lot more. Early in my teaching career, it used to drive me up the wall. I assumed the kid was being defiant or obstinate on purpose... like they just wanted to get under my skin.

But the longer I taught, the more I realized something else: Processing information and organizing your thoughts into something clear is hard. So hard.

It turns out, what I was witnessing wasn't defiance. It was cognitive overload. The student couldn't respond clearly because they hadn't *framed* the task clearly in their own mind. They didn't know where to start, what the question was really asking, or how to organize their thinking. And if you can't frame the problem, it's nearly impossible to solve it.

This is why framing matters.

With AI in the mix, the challenge gets even more layered. Students aren't just responding to a teacher anymore. They're interacting with systems that mirror back whatever they put in... structure, clarity, vagueness, or confusion.

Framing is how we make sense of information, shape our thinking, and direct the tools we use. It shifts the work from answering questions to organizing thought. And whether you realize it or not, you already know how to make this framing shift.

How Framing Supports the SHIFT

By now, we've already done the heavy lifting. We've learned how to observe, describe, evaluate, iterate, and stay in the loop. Framing is the next move that ties it all together.

Whether you're a teacher giving directions or a leader asking AI for help, the way you phrase a question or command shapes the result. Framing adds purpose. It brings clarity. It helps thinking take form.

And in a world filled with generative tools, this skill matters. It helps you survive the AI shift.

We are all learning how to communicate more clearly. With each other. With machines. With ourselves.

Framing supports that process by helping us:

- Add clarity to questions and tasks
- Provide better context and direction for AI tools
- Ask for outputs that are more useful and relevant
- Plan and communicate with greater intention

This isn't about perfect prompts. It's about giving better direction. For your colleagues. For yourself. For the systems we're all learning to navigate.

Crafting the Command

Framing helps you take an idea and turn it into something workable. It starts with a general goal or direction, then moves into the specific pieces that bring that idea to life. This shift from a wide to focused frame transforms vague ideas into clear directions, brings shape to goals, and gives structure to creativity.

This is where the detailed observations honed with tools like the N.O.T.I.C.E. Checklist become invaluable. You can't strategically frame a request to AI with precision if you haven't first deeply seen and noticed the subtle nuances, critical details, or underlying patterns in your data, educational context, or challenge. The quality of your framing is directly tied to the richness of your observation.

These small details matter profoundly. They determine how instructions are given. They change how AI responds. They guide collaboration and make the task more meaningful. By framing our questions and commands with specific information, you give shape to ideas and bring effort to a clear outcome.

How do we make this shift happen?

One of the most practical ways to frame is with a tool that's been around for more than a hundred years: the Kipling Method.

The Kipling Method

One of the simplest ways to frame questions with clarity and purpose is by using the Kipling Method. Inspired by a short poem by Rudyard Kipling, this method uses six familiar question words to guide thinking and structure ideas.

"I keep six honest serving-men
(They taught me all I knew);
Their names are What and Why and When
And How and Where and Who."

These six words (**Who, What, When, Where, Why, and How**) create a flexible, memorable, and powerful framework that helps bring depth, detail, and direction to any prompt or task.

Each one adds a different layer of clarity:

- **Who** is the audience, subject, or key stakeholder (e.g., students, parents, school board, specific staff)?
- **What** needs to be done, created, or addressed (e.g., a lesson, a communication, a policy brief)?
- **When** should it happen, or what time period is relevant (e.g., next semester, for the upcoming school year)?
- **Where** should it take place or apply (e.g., in the classroom, district-wide, for a specific department)?
- **Why** is it important, necessary, or being requested (what's the educational purpose or objective)?
- **How** should it be structured, delivered, or styled (e.g., as a presentation, a concise memo, a playful tone)?

By walking through each of these questions, you can navigate the shift with better commands, clearer prompts, and more useful

AI interactions. It works across grade levels, educational roles, and even outside of AI use. This is just strong communication.

And because the words are already familiar, it's easy to teach and easy to remember.

Modeling the Frame

You get the Kipling Method. Now, let's see it in action. Showing your colleagues how it works helps them see a simple idea turn into a sharp, effective prompt. It teaches them how to shift from a fuzzy question to a focused framing that gets them great AI responses.

Let's start with a common situation: a broad question, maybe the first thing that pops into an educator's mind.

Imagine an educator types: "Tell me about AI."

This first question doesn't give AI much to work with. So, let's see how dropping in those Kipling questions reshapes this basic prompt. It'll become more specific, more useful, and guide AI toward exactly what you need.

Big Picture Framing (Who, What, Why): We can start by adding a purpose and an audience. Asking why you need this info and who you're talking to shifts the whole frame.

- Original Prompt: "Tell me about AI."
- Framed Prompt: "Draft a concise overview of AI for a school board presentation, emphasizing its strategic impact on future education trends."

Detail Framing (What, How): Next, we can add some specific details and say how you want AI to present it. This tells it what to make and how it should look.

- Original Prompt: "Tell me about AI."
- Framed Prompt: "Create a bulleted list of 5 key AI tools relevant to K-12 administration, with a brief explanation of how each could streamline school operations."

Full Kipling Framing (Who, What, When, Where, Why, How): For the sharpest results, throw in all those Kipling questions. This gives AI a fully-framed picture and makes sure it knows exactly what you're aiming for.

- Original Prompt: "Tell me about AI."
- Framed Prompt: "Draft a professional development module outline for elementary school teachers on integrating AI into their daily routines over the next semester, focusing on practical classroom applications and time-saving strategies within our district's technology guidelines."

Let's break down how this powerful prompt uses the Kipling Method:

- **Who:** "for elementary school teachers" (audience)
- **What:** "Draft a professional development module outline... on integrating AI into their daily routines" (core task and subject)
- **When:** "over the next semester" (timeframe)
- **Where:** "within our district's technology guidelines" (specific educational context/location)
- **Why:** "focusing on practical classroom applications and time-saving strategies" (purpose/objective, for effective professional decelopment)

- **How:** "Draft a professional development module outline" (specific format and content requirements)

This is focused and well-framed. It's clear how each layer helps AI deliver a more targeted and useful output. More importantly, it reinforces the systems you've explored so far by leaning into language from the **SEE** stage, keeping a human in command from the **HUMANIZE** stage, and helping you craft questions that deliver your intended results from the **ITERATE** stage.

Let's Frame It!

As you practice framing, remember that this shift is all about adding specific details to a broad idea to transform AI output in a way that makes it much more targeted and useful. The simplest way to practice is to start with a very general question or idea.

Example: "Send an email about a parent-teacher conference."

Next, frame your prompt to expand your request. Challenge yourself to build on that initial idea by adding specific details using the Kipling Method. You should aim to make your prompt far more precise, considering Who, What, When, Where, Why, and How it should be created.

Pause for a moment and consider how you might frame the prompt.

Here is one possibility:
Example (detailed): "Draft a concise, empathetic email to a parent of a third-grade student who is struggling with reading, outlining specific classroom interventions and next steps for

parent support, scheduled for discussion at the upcoming parent-teacher conference."

As an extension to this, you can do a couple different things, like:

Compare the Results: Try both your basic prompt and your detailed, Kipling-framed prompt with an AI tool. Then, compare the two responses and decide which one is better. You can even use a scoring rubric to grade each output.

Label the Prompt: Do this activity in a staff or team meeting. Once you have a collection of framed prompts, swap them between colleagues and have them circle or highlight each part of the Kipling frame with different colors. This allows teams to see several different prompt iterations and also refine their attention to each element of the frame.

In all its forms, these kinds of activities clearly demonstrate how intentional framing leads to much better, more relevant AI output. It highlights that the human's role is to guide AI with clear, detailed instructions.

The Framing Shift

Framing transforms how you think, write, and create. It teaches you to be intentional, to slow down before you press send, and to see prompting as a process of design. This vital skill builds on insights from the **SEE** stage, keeps the human in command from the **HUMANIZE** stage, and ensures every **ITERATION** moves towards the intended outcome.

In a world where anyone can ask a question, the true advantage belongs to those who ask with clarity, context, and purpose.

That's what the **Frame** piece of **The SHIFT System™** is all about.

T: Think

> *"The true sign of intelligence is not knowledge but imagination."* — *Albert Einstein*

My daughter had a homework assignment that I thought was odious. It was nothing but skill and drill... again and again. What made it worse was that I had become the homework police instead of enjoying the little bit of quiet I'd earned that evening.

Eventually, after the bajillionth question, I told her to ask Alexa.

She went over, asked, jotted down the answers, and reported back to me. But when I asked her what the assignment was about, she shrugged and said, "I don't really know."

That moment is exactly why this part of **The SHIFT System™** matters. She had completed the task, but there was no thinking. No meaning. No takeaway. And that's a real risk when AI becomes part of the process. If you only execute tasks or gather information without deeper engagement, you've missed the point.

Everything we've done up to this point by describing with detail, humanizing intent, iterating on drafts, and framing inputs was about setting you up to get better, richer outputs from AI. Now we're at the moment that follows.

This is where we take the response and start to *think.*

Why the SHIFT to Thinking Matters

Even though AI tools can produce answers faster than ever, thinking has value. The ability to think is crucial for all of us stuck in the Sandwich era, quickly moving towards a world that craves creativity more than memorization. We once knew every phone number until smartphones made direct recall less essential. We used to wonder more before search engines provided instant answers. We can't let thinking disappear.

Right now, we have the opportunity to bring back the power of sustained wonder and profound thinking. This final phase is about developing the habits that distinguish human intelligence. It's critical we side-step mindlessly plugging in prompts and copy-pasting outputs. We want to build genuine insight. That's what the THINK mindset is designed to support. When you learn to question a response, not because it's wrong, but because it's worth thinking about, you start to reclaim the cognitive load that makes strategic insight stick.

More than ever, we have to shift to be thinkers.

For those who learn best visually, or seek guided practice to make these shifts easier, The SHIFT System™ Online Accelerator provides in-depth video modules access to the Team Playbook, and dedicated activities. It helps strengthen your SHIFT skills through a guided online experience.

Explore the Accelerator Program at:
SurvivingtheAIShift.com/Accelerator

The THINK Mindset

At this point in The SHIFT System™, you already know how to prompt with clarity, iterate with purpose, and describe your ideas in ways that shape the response. You know how to guide AI toward a better result.

Now the challenge is learning what to do with what you receive.

Surviving the AI Shift means taking what AI generates and using follow-up questions to clarify meaning, examine patterns, challenge assumptions, take action, and reflect on strategic outcomes. These questions create space for you and your colleagues to slow down and think more deeply.

This is where the *work* of thinking begins.

And it's where the human takes the lead.

T – Targeting Questions

When AI responds, it's often just a starting point. You need to learn how to focus in on what's unclear, vague, or overly general. Targeting questions are designed to narrow the output, define terms, and pull the conversation toward clarity. This is where you shift from accepting an answer to making sure you actually understand it.

Example Targeting Questions:

- What does that term mean in this context?
- Can you explain this part using a more specific example?
- What are the key components of this concept?
- How is this different from something that sounds similar?

These questions help you and your colleagues slow down and check your understanding. They create the habit of clarifying before moving forward, which supports deeper comprehension. Thinking starts when you feel confident enough to say, "I don't quite get this yet," and know how to ask a better follow-up.

H – Higher-Order Questions

Sometimes the response makes sense on the surface, but the real work is in connecting it to something bigger. Higher-order questions help you look for relevance, meaning, and relationships between ideas. These questions build the habit of thinking across concepts instead of treating each one in isolation.

Example Higher-Order Questions:

- How does this connect to a previous curriculum initiative or school-wide project we worked on?
- What are the possible effects of applying this idea in a different grade level or school department?
- What pattern or theme shows up when we compare this with other educational institutions or examples?
- What might this reveal about the larger educational issue or institutional system behind it?

Higher-order questions give you and your colleagues a reason to step back and make sense of the bigger picture. This helps you move from surface-level understanding to more complex professional reasoning. The more you connect ideas, the more you start to see insight as something built, not just received.

I – Instigating Questions

This is where thinking gets rebellious. Not every response should be accepted at face value. You need space to challenge what's been presented and think about what might be missing. Using instigating questions introduces a healthy sense of skepticism and invites you to push against easy answers.

Example Instigating Questions:

- What assumptions are built into this explanation or data?
- How might a different educational role or viewpoint see this differently?
- What changes if we remove or reverse one part of this proposed school policy?
- What would a strong counterpoint or alternative solution to this look like?

These questions help you and your colleagues become more aware of bias, perspective, and possibility. You learn that not every answer is fixed or final. Some aren't even factual! This kind of thinking builds independence and flexibility. It also reinforces that questioning is a normal part of the professional practice, not a disruption to it.

N – Next-Step Questions

Some outputs are informative. Others suggest a way forward. But if you stop at reading or understanding, professional progress stalls. Next-step questions help you think about application... how you can use, adapt, or act on the information you've just received.

Example Next-Step Questions:

- What's one way we could try this idea in our current classroom situation or administrative workflow?
- What are some potential obstacles that might come up during implementation?
- What would be the first step in putting this into practice within your school or district?
- What support or resources would help make this educational initiative possible?

These questions help you turn a concept into something useful. They also model a problem-solving mindset. Instead of just understanding what something is, you start thinking about what you can do with it. That shift builds confidence and momentum.

K – Knowledge-Building Questions

When you take a moment to reflect, you often realize you've gained more than just facts. Knowledge-building questions are used to look back, connect ideas, and figure out where to go next. These questions support the long game of professional growth by encouraging synthesis and curiosity.

Example Knowledge-Building Questions:

- How has your strategic thinking changed based on this response or data?
- What questions would you ask next to go deeper into this educational trend or school challenge?
- How does this relate to a past professional learning or institutional experience you've encountered?

- What did this process reveal about how you or your colleagues think or operate?

These questions help you recognize growth. They also show that every conversation with AI can feed into the next one. When you start seeing knowledge as something you're actively building, you stop looking for perfect answers and start thinking in wider, more meaningful ways.

Build a Question Ladder

To help you and your colleagues stretch your thinking after AI gives an answer, start with this:

Give yourself an AI-generated response to a prompt. Then, instead of moving on, you (and your colleagues) walk back through the output with a new lens.

Build a "question ladder" using the THINK model:

- **Targeting**: What needs clarification or better definition?
- **Higher-Order**: What does this connect to or remind us of?
- **Instigating**: What assumptions should we test or push against?
- **Next-Step**: What could we do with this or apply it to?
- **Knowledge-Building**: What new questions does this bring up?

Once you have created one question for each rung, have yourself (or your team) use the ladder to interact with AI again, one question at a time. The goal is to model how deeper thinking leads to stronger, more relevant output.

The Wait-Think Cycle

Just the other day, I was in a deep conversation with an AI chatbot. Deep. It said some things. I said some things... but then it gave me an idea. I knew I needed to get more clarity and ask more questions, but before I did that, I had to stop and walk away. Before I could ask targeted questions, I had to think about my own personal next-step and formulate some connections.

With the wait-think cycle, you simply set up a process in your professional setting where you get AI output and then wait. Yup, wait. Don't ask THINK questions until you have time to think.

In the rush-rush world we live in, we have to take time to wait and think.

When you and your colleagues complete these activities and learn to THINK, the shift is remarkable. The coveted skills we long to cultivate start to flourish. Your engagement is higher, your thinking is deeper, and your professional insights transcend what either human or artificial intelligence could achieve alone.

That's what the **Think** piece of **The SHIFT System™** is all about.

Surviving the SHIFT

This is **The SHIFT System™**: a deliberate move to straddle two mindsets and move from the Information Age to the Intelligence Age without getting lost in the Sandwich Era. By applying **SEE, HUMANIZE, ITERATE, FRAME, and THINK**, you're cultivating the crucial skills that are essential for success in the new

AI-powered world. More importantly, you've developed a consistent, human-centric methodology to leverage AI tools to amplify professional insight, power innovation, and ignite creativity across your educational workforce.

The future of education isn't set by code. It's defined by your actions, by your choices, and by your **SHIFT**. This is how human intelligence evolves.

CHAPTER 4

Leading the AI Evolution

"The future is not set.
There is no fate but what we make for ourselves."
— Kyle Reese, The Terminator (1984)

My husband is convinced that the movie *I, Robot* with Will Smith is a looming prophecy. He genuinely worries we're hurtling toward a future where we ceded too much control, too fast, allowing the machines to "take over." And he's not entirely wrong.

His unease might seem like cinematic paranoia, but the truth is... He might be right. The rapid evolution of artificial intelligence has introduced a host of complex challenges we're only just beginning to grapple with. We're seeing the kinks in the system emerge daily, from fundamental privacy concerns and pervasive implicit biases embedded in algorithms, to thorny issues of copyright infringement and the very nature of truth in a sea of synthetic content. These are real, pressing issues that demand our attention before the future *is* set.

The stakes are high. We simply cannot afford to passively wait to see how things turn out. We have to stay ahead of the curve, before the "robots" (powered by advanced AI) leave us all vulnerable to unintended consequences. That starts by making this profound shift of thought and mindset.

It's imperative we build genuine AI literacy and the critical skills essential for the 21st century. The moment for accelerating professional development across all educators, staff, and administrators is not approaching... It is now.

Time to SHIFT

The good news? We're not helpless. The tools we need to navigate this complex, shifting terrain are right in front of us. It starts with a profound SHIFT of thought and mindset; a deliberate approach to engaging with artificial intelligence that puts human ingenuity firmly in the driver's seat.

This is precisely where **The SHIFT System™** comes in.

Yes, **The SHIFT System™** helps us understand technology and better use AI as a tool. More importantly, it's a strategic guide for developing the kind of AI literacy and critical skills that are truly essential for professional excellence in the 21st century educational landscape. It's how we move beyond apprehension and into empowered action. By consistently applying the principles allowing us to **SEE** details, **HUMANIZE** our interactions, **ITERATE** with purpose, **FRAME** our queries effectively, and **THINK** deeply, you've gained a powerful methodology. Our responsibility now extends to equipping all members of our educational workforce, including the emerging AI-native generations (the Alphas and the Betas), to master these tools and, more importantly, to lead with them.

This skillset cannot be for a select few. We ALL need this across all generations: as educators, as support staff, as administrators, as school leaders, and as individuals navigating this revolutionary moment together.

Surviving the AI Shift means adapting and actively shaping the future. Those futures shouldn't be dictated by algorithms. Instead, they must be collaboratively crafted by human curiosity, ingenuity, and a relentless drive for understanding. We are designing a new reality, and the robots can't hold us hostage because we're building a future where we lead with unparalleled human insight, augmented by the very technology we once feared.

The future isn't set in code. It's defined by your SHIFT.

Welcome to Surviving the AI Shift.

GLOSSARY

AI (Artificial Intelligence): Systems that can perform tasks typically requiring human intelligence, such as learning, problem-solving, and decision-making.

AI-o-Meter: A self-assessment tool (**part of The SHIFT System™**) used to transparently measure and reflect on the degree of AI's contribution versus human input in a task or project.

AI Agents: AI tools designed to perform specific tasks or act autonomously, often interacting directly with users or other systems to fulfill requests (e.g., AI tutors, administrative efficiency agents).

AI-native Generations (Alphas & Betas): Generations born into a world where AI is already ubiquitous and integrated into daily life, fundamentally shaping their expectations for technology and interaction.

Artificial Intelligence Age: The current era characterized by the widespread development and integration of AI, leading to profound shifts in how we work, learn, and live.

Augmented Reality (AR): Technology that overlays digital information or graphics onto the real world (e.g., via smart glasses or phone screens), creating an enhanced view of reality.

Automation: The use of technology to perform tasks with minimal or no human intervention, aiming to increase efficiency and consistency.

BID & ReBID Cycle: A core iterative process (**part of The SHIFT System™**) for engaging with AI, involving an initial **B**rainstorm, **I**dentify, and **D**raft of a prompt (BID), followed by **Re**view and **Ref**inement through repeated cycles (ReBID).

CLEAR Check: A simple, structured framework (**part of The SHIFT System™**) used to critically Check AI outputs for **C**ompleteness, **L**ogic, **E**xplanation, **A**djustment potential, and **R**ewording opportunities.

Descriptive Language: The ability to use precise and rich vocabulary to clearly and accurately communicate observations, ideas, and desired outcomes, especially when interacting with AI.

Dual Lens Exercise: A practical activity (**part of The SHIFT System™**) that trains individuals and teams to skillfully toggle between observing minute details and understanding the broader strategic context, connecting the small picture to the big picture.

Educational Workforce: Refers to all adults working within an educational institution, including teachers, administrators, counselors, and support staff.

FRAME (The SHIFT System™ Element): The principle of defining problems and crafting requests with clarity, purpose, and all necessary context (Who, What, When, Where, Why, How) to ensure meaningful outcomes.

HUMANIZE (The SHIFT System™ Element): The principle of protecting and amplifying the irreplaceable human elements in work—creativity, judgment, empathy, and personal touch—ensuring AI augments, rather than replaces, human value.

Insight Ladder: A collaborative exercise (**part of The SHIFT System™**) that guides teams to move from superficial understanding to deeper insights by asking increasingly complex questions (following the T.H.I.N.K. model).

Intelligence Age: The current era, succeeding the Information Age, where machines can not only access and deliver information but also generate, summarize, analyze, and personalize it in sophisticated ways.

Information Age: The preceding era where access to and mastery of information was the primary driver of value, with technology primarily serving as a tool for storage, retrieval, and transmission.

ITERATE (The SHIFT System™ Element): The principle of continuous refinement and improvement, embracing initial outputs as starting points and relentlessly pursuing "getting it better" through feedback loops and successive attempts.

Kipling Method: A practical framing tool that uses six key questions (Who, What, When, Where, Why, How) to structure thinking and clearly define requests or problems.

Large Language Models (LLMs): A type of AI system (like ChatGPT) trained on vast amounts of text data to understand, generate, and respond to human language.

Multimodal AI: AI tools capable of processing and generating information across multiple formats, such as text, images, video, audio, and code, allowing for richer interactions.

N.O.T.I.C.E. Checklist: A systematic observation tool (**part of The SHIFT System™**) that guides users to look for **N**uances, **O**utliers, **T**rends, **I**nconsistencies, **C**ritical Details, and **E**volution in any data or content.

Professional Practice (in Education): Refers to the daily work, responsibilities, and ongoing development of all adults within educational institutions, including instructional, administrative, and support roles.

Rapid Refine Icebreaker: A quick, dynamic team exercise (**part of The SHIFT System™**) that practices rapid, iterative refinement of communications or ideas under time pressure.

Sandwich Era: A metaphor describing the current transitional period where individuals and institutions are caught between familiar, established ways of working (from the Information

Age) and the rapidly emerging, powerful tools and demands of the Intelligence Age.

SEE (The SHIFT System™ Element): The principle of sharpening one's observation skills to move beyond passive consumption to active, intentional perception of data, contexts, and nuances.

SHIFT (The general shift/transformation): Refers to the profound and accelerating transformation occurring in the world due to artificial intelligence, requiring adaptation in how we work, learn, and live.

The SHIFT System™: A proprietary methodology for increasing AI literacy, built upon five core principles: See, Humanize, Iterate, Frame, and Think.

Strategic Insight: Deep understanding derived from critical analysis and purposeful questioning, leading to actionable wisdom and informed decision making.

T.H.I.N.K. Question Compass: A systematic framework (**part of The SHIFT System™**) for asking targeted, higher-order, instigating, next-step, and knowledge-building questions to extract deeper insights from information or AI outputs.

THINK (The SHIFT System™ Element): The principle of engaging in profound critical inquiry, challenging assumptions, and transforming raw data or AI outputs into actionable wisdom and strategic advantage.

Wait-Think Strategic Pause: A practice (**part of The SHIFT System™**) of deliberately pausing after receiving information (especially from AI or in high-pressure situations) to allow for deeper, unpressured strategic reflection before acting.

WonDAR Framework: A simple, human-centered framework **(part of The SHIFT System™)** for interacting with AI, emphasizing **W**onder, **D**irect, **A**dapt, and **R**eview to maintain human agency and strategic thinking.

ABOUT THE AUTHOR

AI is reshaping our world, and I believe the future of education hinges on how well adults within the system adapt and lead. My driving philosophy is **Forward Ever, Backward Never**. This belief has fueled my entire career.

For two decades, I've been right there in the trenches, from inner-city classrooms to international consulting, helping people untangle complex ideas and get truly good at new ways of working and thinking. My focus has always been on making big shifts feel human-friendly and genuinely useful. I've seen firsthand how clear thinking and adaptability open up new opportunities.

As an AI strategist and professional development expert, my passion is ensuring that every educator, every counselor, every administrator, and every staff member can confidently embrace the Intelligence Age. That's why I built **The SHIFT System™**, a straightforward, step-by-step approach to help educational organizations cut through the AI noise and get truly strategic, by showing their teams how to think critically with AI, collaborate effectively, and simply get smarter results.

I'm Ayo Jones, M.Ed. and as an author of *Surviving the AI SHiFT* and creator of *The SHIFT System*™, I inspire leaders and their teams to get ready for this new era. After filming the HBO Max docuseries, 'Coming From America,' I took that inspiration global from my home in Ghana, West Africa.

If you're a superintendent, an L&D director, or an education leader aiming to equip your entire staff to lead powerfully with AI, cultivating sharper, more strategic thinkers, let's connect. We have vital work to do together.

Ready to implement The SHIFT System™ across your entire organization?
We offer comprehensive solutions tailored for districts, colleges, and charter schools:

- **AI Literacy Jumpstart:** Foundational training for all adult staff.
- **Online Accelerator Program:** Deeper, structured online learning.
- **Certified Educator Facilitator Program:** Build internal capacity to scale AI expertise.
- **Direct Leadership & Strategic Partnership:** Personalized advisory and high-impact engagements.

Let's discuss how we can create a custom solution for your school or district.

Visit SurvivingtheAIShift.com or email us at: Hello@SurvivingTheAIShift.com.

www.ingramcontent.com/pod-product-compliance
Lightning Source LLC
Chambersburg PA
CBHW071534120626
46550CB00006B/2460